Ralph E. Flanders

Construction and Manufacture of Automobiles

(1912)

Ralph E. Flanders

Construction and Manufacture of Automobiles

(1912)

ISBN/EAN: 9783845713120

Erscheinungsjahr: 2011

Erscheinungsort: Bremen, Deutschland

www.unikum-verlag.de | office@unikum-verlag.de

Ralph E. Flanders

Construction and Manufacture of Automobiles

(1912)

CONSTRUCTION AND MANUFACTURE OF AUTOMOBILES

By Ralph E. Flanders

CONTENTS

Design and Construction of a High-grade Motor Car - 3
Automobile Manufacturing Methods - - - - 18
Manufacturing Equalizing Gears - - - - - - 31

CHAPTER I

DESIGN AND CONSTRUCTION OF A HIGH-GRADE MOTOR CAR*

The following description of a 40 H. P. automobile, built by the Stevens-Duryea Company, of Chicopee Falls, Mass., may, except for certain important details which will be specifically mentioned, be taken as typical of the design of high-grade cars in general. In Fig. 1 is shown a side view of the "Model Y," 40 horsepower, six-cylinder machine, with 36-inch wheels and 142-inch wheel-base. An automobile may be divided into two parts—the body and the "chassis." The former is the product of the carriage-maker's art, the latter of the mechanic's

Fig. 1. Stevens-Duryea "Big Six" Motor Car, 1910 Model

and engineer's. The chassis of this machine is shown in Figs. 2 and 3, to which reference will now be made.

The mechanism and body of the car are supported by a frame whose side members, of chrome-nickel steel, are shown at *A*. These are connected by four cross pieces, and are supported on the front and rear axles by the spring connections shown. The cross pieces are also pressed from chrome-nickel steel, and are hydraulically riveted to the side frames. A platform spring suspension is used at the rear, hung on connecting shackles designed to overcome the side roll met

* MACHINERY, October, 1909.

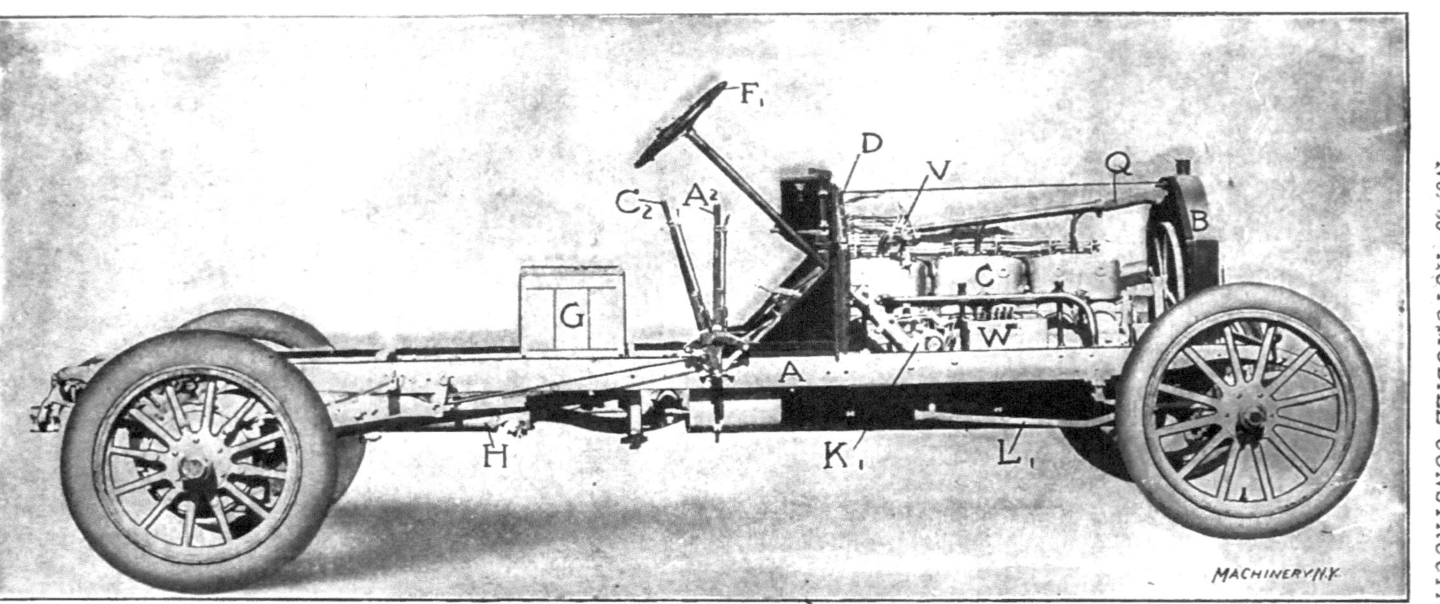

Fig. 2. Side View of Car with Body Removed, showing Chassis

when rounding curves in large and fast cars. The springs are made from steel selected after careful tests of both American and imported materials. The cost of the brand selected was far in excess of that of the nearest competitor, but it gave an endurance under repeated shock and reversal of stress not met with in any other make.

On the chassis frame are mounted, first the radiator *B*, next the engine C, then the dash-board *D* with its steering and controlling mechanism, the clutch and speed change mechanisms at *E* and *F* respectively, the gasoline tank *G*, the muffler *H* for the exhaust, the propeller shaft *J* for transmitting the power to the rear axle, and the rear axle

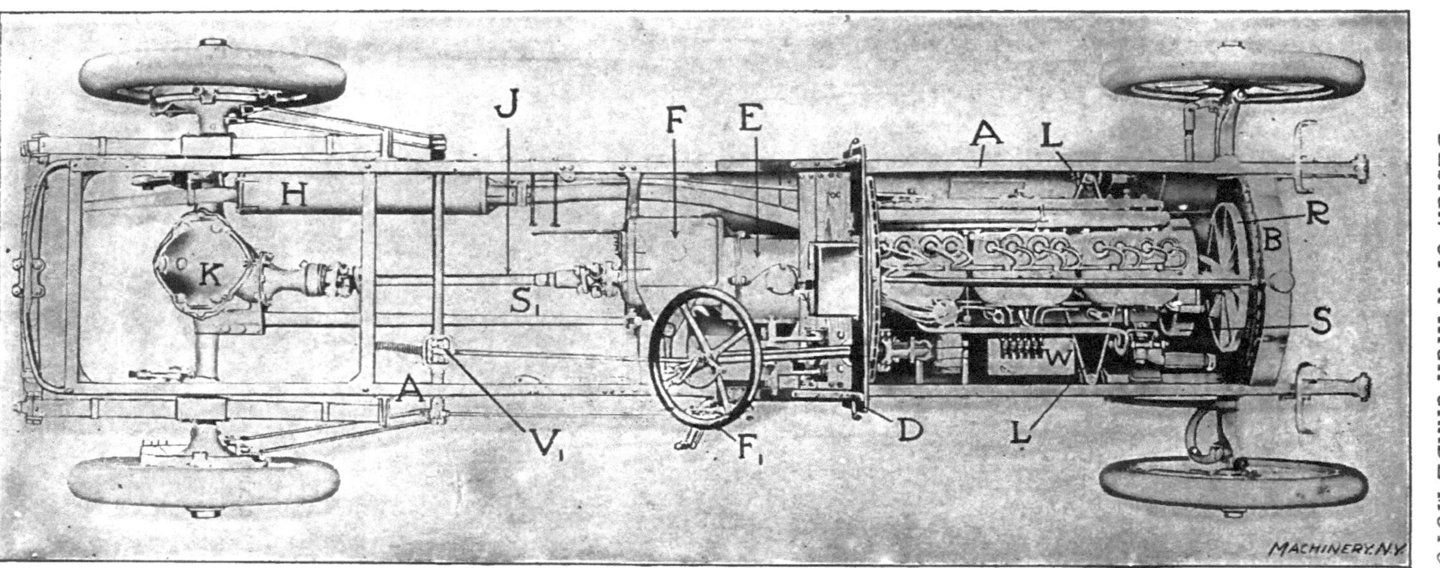

Fig. 3. Top View of Chassis, showing Arrangement of the Mechanism

with its differential gearing at *K*.

The engine is shown more clearly in Figs. 4 and 5, which show the "unit power plant" form of construction, one of the important original features of the design. This peculiarity consists in mounting the engine,, clutch, and transmission casings as a single rigid member, supported by a three-point bearing on the flexible frame. Supports *L* bear on the two side frames, while pivot *M* is riveted to one of the cross pieces. This allows the whole of the contained mechanism to run without distortion or bending, even on roads which rack the frame severely, and thus results in less friction and lighter structural parts, giving a high

Fig. 4. The Unit Power Plant, comprising Engine, Transmission, etc.; View taken from the Timer and Lubricator Side

available horsepower per hundredweight of load. It also permits the power plant to be assembled as a whole and to be bolted in place without fitting. This construction, which is the distinctive point in the design of this motor, has been successfully followed by the builders for the last five years, and it is one of the things which serve to give an attractive mechanical appearance to the whole mechanism. Only one double set of universal joints is required, that connecting the propeller shaft with the transmission gearing at one end, and the differential gearing at the other.

The cylinders are grouped in three two-cylinder castings *C*, bolted to the crank case *N*. As is common with internal combustion engines in ordinary practice, they are water jacketed, there being a continuous

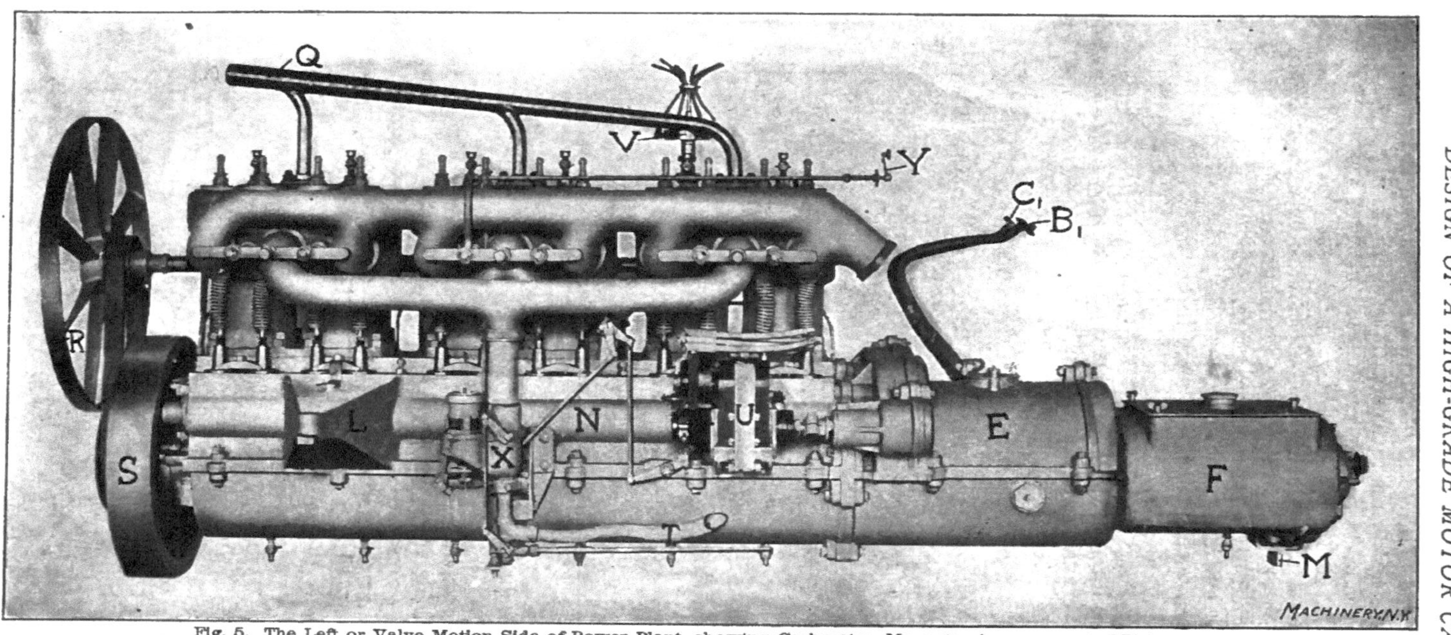

Fig. 5. The Left or Valve Motion Side of Power Plant, showing Carburetor, Magneto, Arrangement of Piping, etc.

circulation from radiator B through centrifugal pump O and pipe P to the water jackets, thence back again through the return pipe Q to the top of radiator B. Here the heated water is cooled by passing through sheet metal channels, having a large radiating surface exposed to the draft of wind produced by the passage of the machine through the air. This draft is increased by an aluminum fan R belted to the pulley on the outside of flywheel S. An automatic tightening arrangement is provided for the belt.

It should be mentioned that the placing of the fly-wheel

at the forward end of the crank-shaft, as here shown, is unusual, the common construction being to locate it between the crank-shaft and the clutch. It tends, in particular, to bring more of the weight onto the front wheels, off from the heavily loaded rear wheels of the machine, and permits the reducing of the clearance over the roadbed in the center of the chassis, where there is the greatest danger of striking on high water-bars, railroad crossings, etc. It will be readily seen that more clearance is required at the center of the machine than at the axles, when crossing a hump in the road.

Lubrication, Ignition, etc.

Two shafts mounted in the crank casing, one on each side, above and parallel to the crank-shaft, are driven from it by enclosed gearing.

Fig. 6. View of Engine from Beneath, showing Removal of Piston, Cam- and Lay-shafts, etc., without Dismantling

The one at the side shown in Fig. 5 is the cam-shaft and is provided with twelve sets of cams for operating the six inlet and six exhaust valves, whose stems and closing springs are plainly shown in the engraving. The driving gear of this cam-shaft is also connected with a pinion on the armature shaft of the magneto, whose function will be described later. The shaft on that side of the machine shown in Fig. 4, is known as the lay-shaft. Its office is the driving of the timer *V*, which controls the ignition, the driving of the forced lubrication mechanism at *W*, and of the water jacket circulation pump *O*.

The lubricator gives a forced oil supply with sight feed, and is always in operation when the engine is in motion. The six-throw crank-shaft is mounted in four bearings in the crank case, with two cranks between each pair of bearings. The boxes at these points are connected with the lubricator *W*. The lower half of the crank case forms a reservoir for the oil escaping from the main bearings. The connecting-rod splashes into this and thus supplies the pistons, connecting-rod bearings, etc., with the necessary lubrication.

The ignition in each cylinder is effected by either of two systems, the one by storage or dry battery and induction coil, and the other by means of a magneto *U* connected by gearing with the crank-shaft. The battery and spark coil is used in starting, while the magneto is used for regular running. The spark coils and switches are located on the dashboard. A lever on the steering wheel, as will be described, is connected with the commutator or timer *V*, which distributes the current to the six cylinders in such a way as to enable the operator to advance or retard the spark at will.

The Carburetor and Fuel Supply

An important and rather delicate piece of apparatus essential to the operation of the gasoline engine, is the carburetor, shown at *X* in

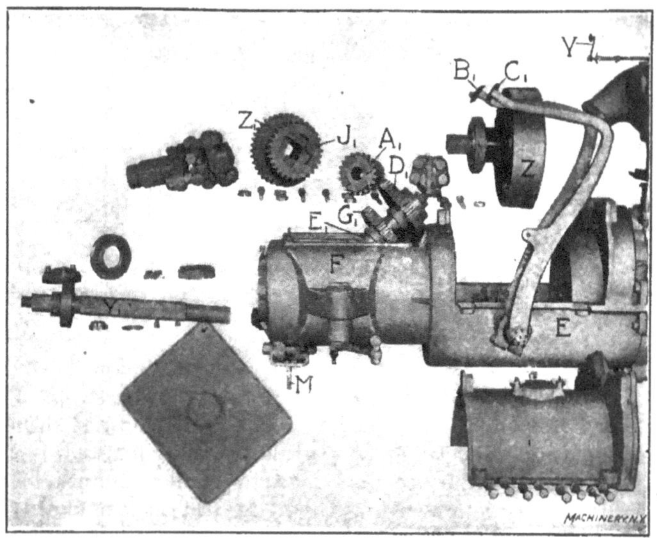

Fig. 7. Clutch and Transmission Gear Members Dismantled to show Construction

Fig. 5. This receives a supply of gasoline through a feed pipe from the tank *G* (see Fig. 2), a supply of air through *T* heated by the exhaust gas for vaporizing the gasoline, and a supply of fresh air to furnish the oxygen for the charge. The gasoline is received in a float chamber, where the level of the liquid is maintained by a suitable float and valve. An automatic valve provides for a constant proportion of oxygen and fuel at widely-varying speeds. The carburetor is provided with a throttle which controls the needle valve connection in the feed pipe, together with the butterfly valve in the suction to the cylinders, thus providing the driver with means for varying the amount of charge furnished the machine; this controls the speed without shifting the gears in the transmission case. The automatic air valve is controlled from the seat by a handle *Y* on the dash-board, which permits the obtaining of a proper mixture for the starting. A button at the front of the radiator, where the machine is cranked for starting, also provides means for flooding the carburetor with fuel for a send-off. The throttle is controlled from a lever on the steering wheel, concentric with the spark control lever, or from an "accelerator pedal" on the foot-board.

The gasoline supply tank *G* is located under the front seat. It contains a partition near the bottom which saves about three gallons out of its twenty gallons' capacity, for use in emergency. By the manipulation of cut-off valves passing through the left side frame of the chassis, it is possible to use this reserve supply after the tank has been otherwise exhausted. This provision is a great comfort to the motorist at critical times.

The Clutch and the Transmission Gearing

In casing *E* is mounted the clutch *Z* (Fig. 7) connecting the engine with the transmission to the driving wheels. This is of the multiple

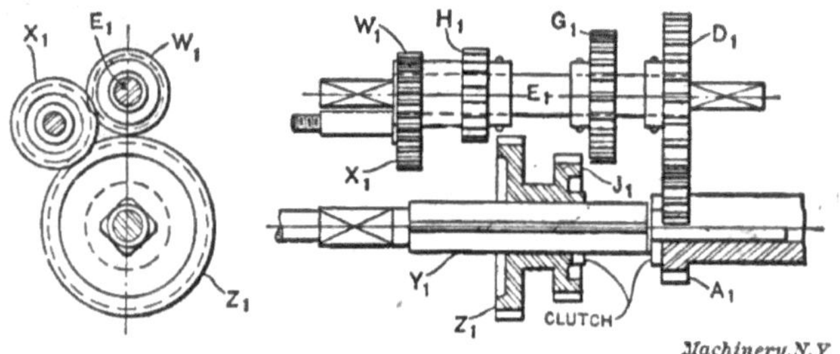

Fig. 8. Sketch showing Arrangement of Gears in Transmission Case

disk type, with alternate disks keyed to the driving and driven members. The driving disks have a wired asbestos facing which makes a superior friction surface, and gives a high resistance to heat as well. This construction obviates, and in fact makes impossible, the use of oil in the clutch. The friction surfaces are held in engagement by a spring, and are released by a pedal B_1, which projects through the foot board at the driver's side of the machine. The spring is so proportioned as to give a smooth, easy engagement, entirely out of the control of the driver, who thus finds it impossible to start the machine with a sudden shock. The second foot lever, C_1, is connected with the rear wheel brakes, as will be described. The driven member of the clutch is connected with the driving shaft in the transmission case or speed box *F*. Contained within it is a mechanism which, by the aid of the sliding gears, clutches, etc., permits of the obtaining of three forward and one reverse speed.

The operation of this gearing will be understood from the sketch shown in Fig. 8. Gear A_1 receives its movement from the clutch. It meshes with gear D_1 keyed to the secondary shaft E_1, which is thus in motion whenever the engine is running and the clutch is engaged. This shaft carries also gears G_1, H_1, and W_1, the latter of which drives, in turn, the idler X_1. Squared shaft Y_1 is directly connected by means of propeller shaft J (Fig. 3) and the universal joints with the rear axle. On Y_1 is mounted the double sliding gear Z_1. Clutch teeth are provided in the faces of the gears A_1 and J_1.

In the position shown in Fig. 8, the transmission is in the neutral position, so that the motion from the clutch is not transmitted to the axle. The right-hand end of shaft Y_1 lies loosely in the revolving gear A_1. When the sliding gear is thrown to the extreme right, the clutch faces of A_1 and J_1 are engaged, so that shaft Y_1 is driven directly. and at the highest speed, from the clutch. By shifting it a step to

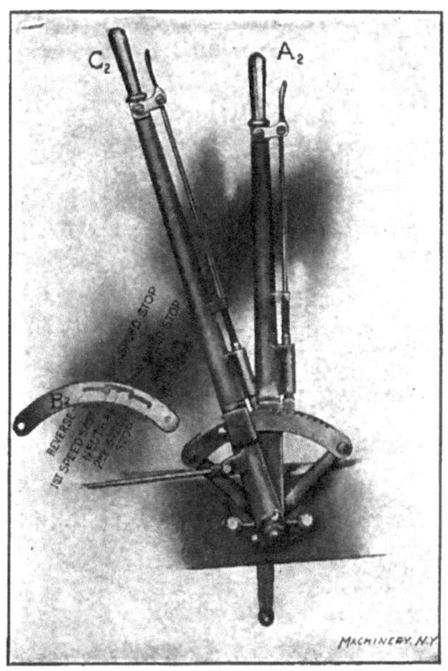

Fig. 9. The Speed Gear Control and Emergency Brake Levers

the left, J_1 is thrown into mesh with G_1, thus giving a lower rate of speed through the back gear shaft E_1. A still further movement to the left, past the neutral point shown in the engraving, brings Z_1 into engagement with H_1, giving the lowest forward speed. A final movement to the left engages Z_1 with idler X_1, thus reversing the drive.

The shifting of gears Z_1 and J_1 is effected by a forked lever connected with lever A_2 (Fig. 9) at the side of the machine, which thus controls the speed changes. This lever is provided with a latch connected with a pin in the slot of the quadrant B_2, operating in a manner easily understood from the engraving. It will be seen that it is possible to move between the reverse and the lowest speed, or between the second and the high speed, without touching the latch, and it is possible to make all the movements rapidly and precisely by the sense of touch without looking at the quadrant at all.

The Differential Drive

Propeller shaft J leads from the transmission case F to differential case K on the rear axle. The bevel gear M_1 (Fig. 11) is connected with the two rear wheels by a differential mechanism, whose function it is to give an equal tractive force to each of the two wheels, but at the same time to permit either of them to run ahead or lag behind the other as may be required in rounding curves, riding over obstructions, etc. The principle of this mechanical movement will be understood by referring to Fig. 10.

Referring first to the sketch at the left, N_1 is the pinion on the propeller shaft and M_1 is the driven bevel gear, concentric with the axle.

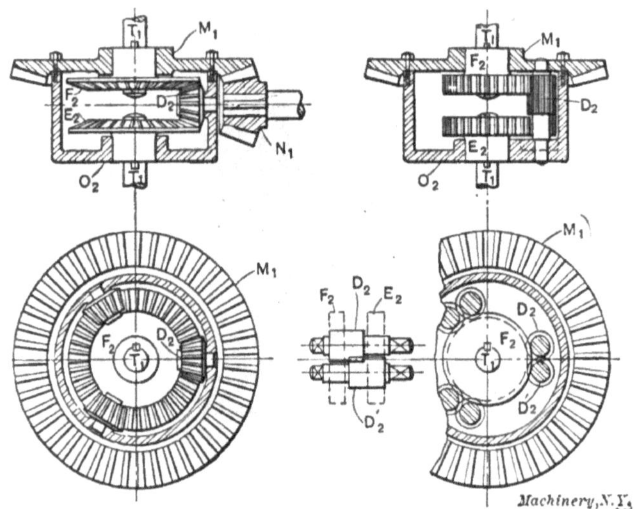

Fig. 10. Sketch showing Principle of the Bevel and Spur Gear Types of Differential Gearing

This gear and shell O_2 to which it is bolted, revolve freely on the hubs of E_2 and F_2. Within the shell are mounted radial pivots on which revolve, loosely, bevel pinions D_2. These engage with bevel gears E_2 and F_2, connected respectively with the right- and left-hand axle shafts T_1. It will be seen that under ordinary conditions the rotating of gear M_1 carries gears E_2 and F_2 along with it, by the pull exerted on them by the bevel pinions D_2, which are stationary; thus the two rear wheels are driven at the same rate of speed. Suppose now that the right-hand wheel be held from turning, so that gear E_2 is stationary, then the rotation of bevel gear M_1 will roll pinion D_2 about on E_2 with a compound action, which will give F_2 twice the rate of speed it had before. In the same way, F_2 can be held from revolving, in which case E_2 will have twice its normal speed, or either of them may be slowed down, in which case the other is speeded up correspondingly. The driving force on both wheels, however, is always the same.

An alternative form of this device is shown at the right of Fig. 10, in which each of the bevel gears D_2 is replaced by a pair of spur pinions D_2 and D'_2, meshing with each other and with spur gears E_2 and F_2 as shown. A little study will show that the action of this device is identical with that shown in the sketch at the left of the figure, the only change being the employment of spur gearing in place of bevel gearing. The differential used on the Stevens-Duryea machine is of the second or spur gear type.

The Full Floating Type Rear Axle

The differential gearing is contained in the casing O_1, which forms the central member of the axle. Tubular extensions to both sides

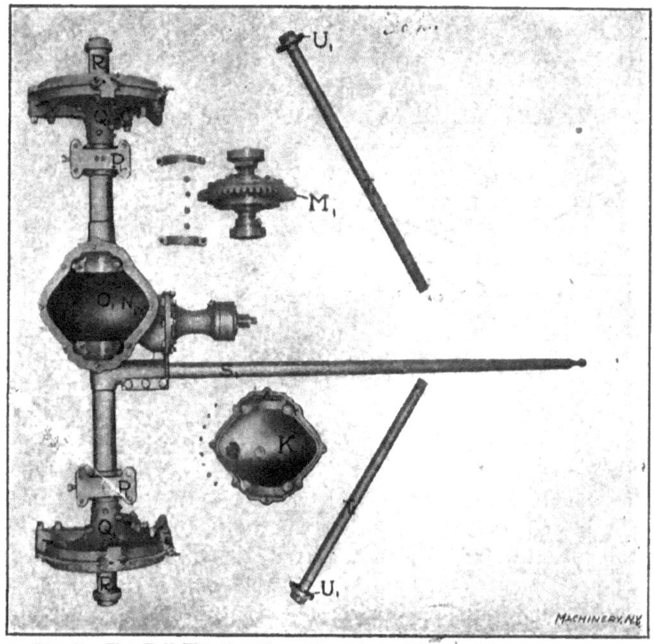

Fig. 11. The Full Floating Type Rear Axle, Differential Gearing, etc.

carry the spring supports P_1 on which the weight of the car rests. The brake flanges Q_1 and the wheel bearings at R_1, all of which are solid with each other, are non-rotating. The rear axle, however, is permitted to rock in spring supports P_1. The torque rod or tube S_1, which is fast in case O_1, extends toward the center of the chassis, where it is hung in a spring suspension as seen in Fig. 3 permitting a limited vibration up or down, with a constant force urging it toward a central position. This construction furnishes the resistance against the climbing of pinion N_1 on bevel gear M_1. In case of sudden starting or stopping, a limited amount of climbing either way is permitted, the torque rod being raised or lowered against the spring pressure to correspond. This greatly decreases the danger of gear breakage.

The construction just described belongs to what is known as the full floating type axle. The wheels are mounted on ball bearings on stationary journals R_1. Shafts T_1 are provided with squared driving ends engaging sockets in the differential gearing in casing O_1 at one end, and similar sockets cut in driving dogs U_1 at the other end. These latter members have driving slots engaging dove-tails in the hubs of the wheels, to which the power is thus transmitted. The squared ends of shafts T_1 are rounded to permit a slight rocking movement in their sockets in the differential gearing and driving dogs U_1. This permits the springing of the rear axle under the load without cramping the driving mechanism.

To allow for the springing of this axle under the load, the two sections of tubing on either side, between members O_1 and Q_1 are held in bored seats which point downward at an angle of ½ degree from the horizontal on each side. Thus the rear axle wheels point in toward each other at the bottom at an angle of ½ degree from the vertical, giving a much better appearance than would be the case if they should by some mischance point the other way. It would take a load in excess of any which would ever be applied to spring the axle and bring the wheels into the vertical plane. It is stated that when the wheels are exactly vertical, they have the appearance of being sprung out at the bottom, into the position occasionally seen in a vehicle of the "one-horse-shay" type.

The Brakes

The brake mechanism of the automobile is of the utmost importance, as is realized by anyone who has had anything to do with these machines whether as driver, passenger or pedestrian. It is usual to provide two complete sets of braking machanism, one for regular use and the other for emergency. That for regular use is controlled by the foot lever C_1 (see Fig. 4), which is connected with a reach rod leading to double cranks on a transverse rock-shaft at V_1 (Fig. 3). One section of this rock-shaft is connected with the brake at the right side of the machine, and the other at the left. An equalizing lever between the two insures an even pressure on each of these two brakes, even though one be much more worn than the other. The brake is of the band type, applied to the outside of a brake rim fast to the hub of the wheel. The emergency brake is operated by lever C_2 (Fig. 2). This, by means of a second rock-shaft concentric with V_1, controls internal expanding ring brakes in the hubs of the wheels.

The Control of the Machine

The steering gear will be best understood from Figs. 2, 3 and 12. The wheel F_1 is mounted on a tubular shaft which carries at its lower end a worm engaging the segment of a worm-wheel G_2 in casing K_1. To the hub of this segment is connected a bell crank H_2 which, through the operation of the steering rod L_1 (see Fig. 2) and suitable connecting cranks and links, turns the front wheels to the right or left as may be required. Spring cushions are provided at the ends of steering rod L_1 so that sudden shocks and twists of the wheels are not transmitted to the worm-gearing and the steering wheel, even when traveling at a high rate of speed. As most mechanics doubtless know, the center line of the pivots about which the wheels are swiveled meets the road at about the point where the tire touches it. This makes it possible to turn the wheels easily when standing still, and decreases the danger of accident while running, as well.

As previously stated, the throttle control and the timing of the spark are effected frcm levers placed at the hub of the steering wheel.

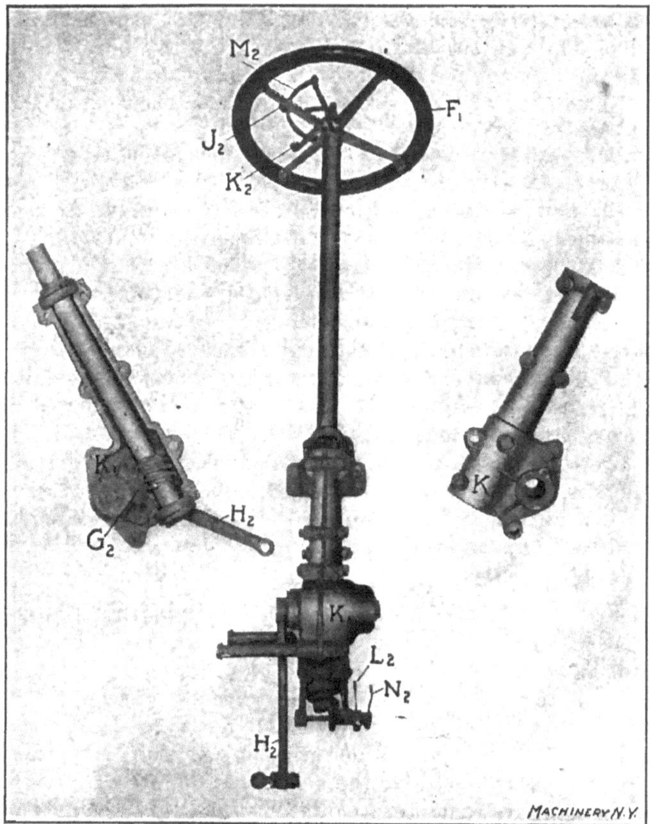

Fig. 12. The Steering Post, with its Throttle and Sparking Connections

Lever K_2 controls the throttle. This is mounted on a tube passing through the steering wheel tube and connected at its lower end by bevel gear segments with a bell crank L_2, which is, in turn, connected by suitable rods and levers with the carburetor. Inside of the throttle lever tube is still another fixed tube on which is mounted the segment M_2, which is thus held stationary. This is provided with notches for locating lever K_2, and lever J_2 as well, which latter controls the timing of the spark. This is mounted on a rod which passes through the center of the system of tubes and is connected by bevel segments with lever N_2 leading to the commutator or timer V.

It may be well to recapitulate as to the functions of the levers, etc., used in the control of the machine. At the front of the radiator is the crank by which the motor is turned, for starting. By the side of it is a button connected with the carburetor, for flooding the latter at starting to obtain a rich mixture on the first stroke. On the dashboard is mounted a lever Y, for setting the automatic air valve to supply the proper amount of oxygen for starting. Beside it is a switch for throwing the ignition spark from the battery to the magneto when the machine is changed from the starting to the running condition, and *vice versa.* On the dashboard are also mounted the spark coils. Through the foot board project the two pedals B_1 and C_1 controlling the clutch and the operating brake respectively, as described. Hand lever C_2 and A_2 control the emergency brake and the speed changes respectively.

Two small pedals are also provided on the foot board. One of these is connected with the throttle in such a way that this may be controlled by the foot instead of by the hand if required. It is called the accelerator. By its use, when the hand throttle lever has been set to a certain point, the valve may be opened clear out to the maximum, as desired, by the foot, thus giving immediate control under varying conditions of traffic. The other pedal operates a valve which cuts out the muffler. This is occasionally done to make the exhaust audible, for finding out how the engines are working, and also for removing the back pressure, and thus giving every ounce of power possible on critical occasions.

These levers, pedals, etc., with the main and supplementary gasoline supply valves previously mentioned, give the driver complete control of a powerful, swift machine, if he has the knowledge, experience and nerve to use them properly.

General Considerations in Automobile Design

A glance at the illustrations will serve to show that the chassis of the modern high-power automobile is a rather complicated, highly specialized, and carefully designed piece of mechanism. It is within the memory of the child in kindergarten when this was not the case, and the writer has painful memories of his duties as consulting physician to one of the best of the machines in existence six years ago At that time, the mechanism of the automobile did not have the homogeneous, appropriate structure that the successful machines of the present day possess. It had a gasoline engine, an epicyclic speed change mechanism, a jack-in-the-box differential gear, and chains leading to the rear wheels of a "horseless carriage." Over the mechanism thus described wandered a maze of levers, braces, pipes, wires, etc., supported at intervals at any part of the mechanism which happened to be in convenient reach. That, however, was before the automobile "found itself." The present development has been the result of the experience of many men with break-downs and failures, as well as of an enormous amount of theoretical work in the matter of testing of materials and analysis of conditions. These theoretical and practical results have been combined on the drawing board, and the resulting machine has the appearance of having been *designed* rather than simply *built*.

The guiding principles in the design of the automobile relate to strength, power, lightness, durability, accessibility, and economy in operation. The matter of economy in construction and materials is about the last thing to be thought of, instead of the first, as with many other classes of machinery. The severe and often reckless usage received by one of these machines demands special treatment in the design and construction which should not ordinarily be necessary.

As an illustration of what has been said in this respect, attention may be called to the method of connecting the driving members of this machine, from the engine through to the wheels. In no place throughout the length of the chassis are keys used for this work. Reliance is everywhere placed on square joints or dovetailed flanges. The crank-shaft is connected with the driving member of the clutch by a square taper socket. The driving member of the clutch is connected by a square socket with the driving shaft of the transmission gearing. The sliding gears of this mechanism are mounted on square shafts, and the same squared drive is used for the universal joints, propeller shafts, pinion shafts, etc., through the intermediate pinions in the differential gearing at M_1 in Fig. 11, and through driving shafts T_1, to the driving dogs on the wheel hubs. These latter, as well as the side plates of the differential gearing, drive or are driven by the engagement of dovetailed teeth. The possibility of the shearing of keys, always present in machine parts subject to shock, is thus avoided. The makers believe themselves to be the only firm employing a complete drive of this kind.

In the matter of accessibility, a study of Figs. 6 and 7 will be found interesting. By removing the lower crank chamber casing and turning the crank-shaft to the proper position, the piston and piston rod may be removed without further trouble, and without removing cylinders or cylinder heads. The same is true of the cam- and lay-shafts. The covers provided for the clutch and transmission casings give evidence of care in providing easy means for inspection and removal of all parts likely to need attention. With a well-designed machine the man on his back under the motor car is a mere figment of the imagination.

CHAPTER II

AUTOMOBILE MANUFACTURING METHODS*

The subserviency of manufacturing considerations to considerations of strength, durability, accessibility, etc., mentioned in the preceding chapter, results in the design of parts which require special and interesting provisions for their economical production. Only a few of the operations particularly noticed in the Stevens-Duryea factory will be described here. They will serve, however, to give an idea of the general practice in such work, and will illustrate the ingenuity required for the solution of some of the problems.

Operations in the Machining of Cylinders

In Fig. 13 is shown a Beaman & Smith combined horizontal and vertical milling machine engaged in surfacing the base, exhaust and

Fig. 13. Gang Milling Operation. Surfacing Cylinder Sides and Ends

inlet flanges, and the spark plug bosses of a series of cylinder castings. The work is mounted in gangs according to the most approved methods. The picture is chiefly interesting in that it shows that the builders take advantage of wholesale manufacturing methods even in the building of a $4,000 machine. Of course, an extensive use of jigs and fixtures, besides reducing the cost of manufacture, results in a greater uniformity in the product, and thus gives the advantage of an easy renewal of worn or damaged parts.

Fig. 14 shows a Beaman & Smith boring machine with fixtures mounted on the rotating table for holding four double cylinder castings. This table can be rotated and adjusted across the bed of the machine.

* MACHINERY, October, 1909.

On each side of the table, double boring heads may be fed in along the bed, one carrying roughing and the other finishing cutters, the feeds and speeds of the two heads being independent. A set of two castings being in place on the roughing end, the head is fed into them and one hole in each casting is roughed out. The work-table is

Fig. 14. Four-cylinder Boring Machine with Revolving Table

Fig. 15. Grinding the Cylinders. Note Connections for Exhausting the Dust and the Use of the Water Jacket for Cooling

then shifted, by means of the hand-wheel, against suitable stops, and the other bore of each cylinder is roughed. The table is then indexed to bring these castings to the finishing side, where the same operation is repeated, the boring being here carried to size for grinding. This rotating of the table, in turn, brings a new set of the cylinders up to be rough-bored. The process is continuous, the work being removed from the finishing side and new cylinders clamped in, while the rough boring is being completed.

For setting out the cutters in the boring bars, the construction shown in Fig. 16, at the left, is used. It will be seen that a taper-headed screw is used for forcing the blades out simultaneously. The cutters *B* bottom on this taper-headed screw *C*; fillister head screws *D* serve to keep the blades forced down to their bearing on *C*, and so draw them firmly against the side of the slot. By this means two or more blades may be set out simultaneously for regrinding to exact size. A similar arrangement (see view at the right of Fig. 16) is used for cutters in the middle of long boring bars, except that the taper point of a screw tapped into the bar from the side, is used in place of the corresponding taper-headed screw in the first case.

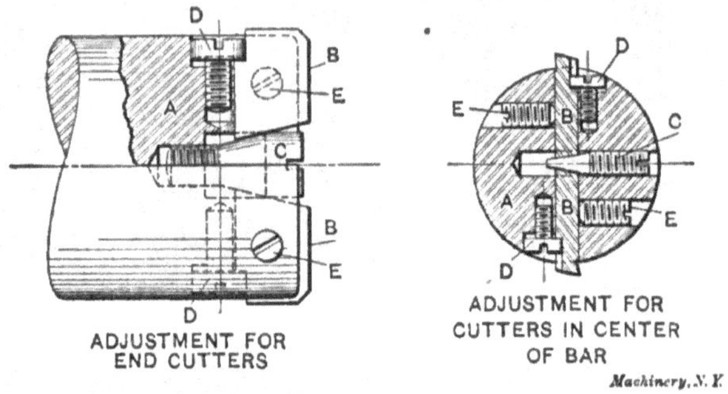

Fig. 16. Adjustment used for Boring-bar Blades

The bore of these cylinders is finished in Heald internal grinding machines especially built for this work. These are of the type in which the work remains stationary while the axis of the spindle is revolved about the center line of the bore and parallel with it, on such a diameter as to bring the outer periphery of the wheel in contact with the inner surface of the bore. The grinding spindle is fed out so as to rotate in a larger circle as the diameter of the bore is increased. An interesting feature shown in Fig. 15 is the provision of a flexible suction tube for drawing out the dust of the grinding through the inlet and exhaust ports, and also the provision made for water cooling. The water is not applied directly to the wheel, as in an ordinary external grinder, but is forced instead through the regular water jacket of the cylinder casting. This reproduces, in a measure, the conditions met with in actual use, and so tends toward accurate work.

Machines and Fixtures for Grinding and Lapping

There are other operations of interest in the grinding department besides that of finishing the bore of the cylinders. Extensive use is made of the Pratt & Whitney face grinding machine for finishing flat surfaces; in fact, it has largely displaced the vertical milling machine for this work, on parts in which the surface to be finished is clear of projections or obstructions to the sweep of the wheel. The faces of the various casings, covers, inlet and exhaust pipes, etc., are finished on this machine. In the past most of these parts have been made from castings on which 3-16-inch of stock had been left, in accordance with the usual practice of milling. The castings come true enough to shape, however, to permit of this finish being reduced to 1-16 of an inch, or

Fig. 17. The Acme of Simplicity in Fixture Making. Face Grinding the Steering Gear Casing

Fig. 18. Grinding the Bore of the Cams Concentric with the Cylindrical Surface

thereabout, when finished by grinding, thus materially reducing the time required. Even when removing 3-16-inch of stock the grinding machine has proved its superiority to the milling machine in the matters of cost, finish and accuracy. The foreman of the grinding department discovered that a little experimenting and investigating along the line of the grading of wheels made a tremendous difference in their durability and effectiveness in removing metal. For aluminum work a vitrified carborundum wheel of about No. 24 grain and grade H hardness is used, a soda compound being employed for cooling.

The cover side of the steering gear casing is one of the parts surfaced on the face grinder. An exceedingly simple fixture is used for holding it. This fixture, as may be seen in Fig. 17, is nothing more or less than a mass of lead melted and poured around a sample casting as a form. The work is set into the bed, thus prepared to receive it, and is supported on the table by its own weight, no fastening being necessary. The castings come uniform enough so that they fit well in this device, except at certain points around the gates and sprues, where it is found necessary to relieve the form slightly to allow for these variations. It may be mentioned that the other or main member of the steering gear casing has a boss projecting above the finished

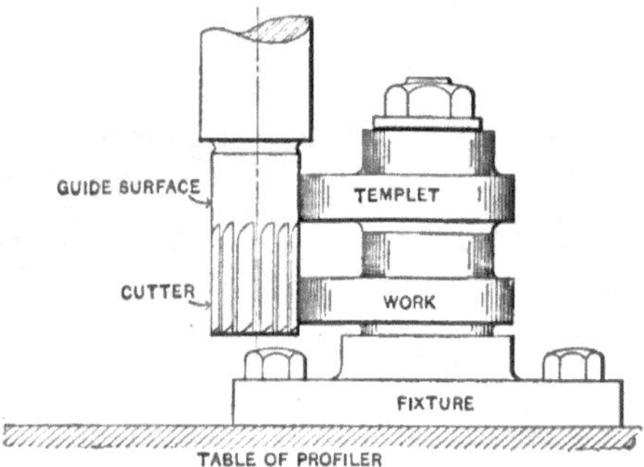

Fig. 19. The Simplest and Stiffest Arrangement for Cam Cutting

surface of the joint, making it necessary to mill that surface. The joint is thus formed of one ground and one milled surface.

In Fig. 18 is shown the operation of grinding the holes in the cams. It is quite important that the cylindrical portion of the cam shall be exactly concentric with the cam-shaft to prevent shock or jar during the period when the valves are supposed to be closed. To make sure that this surface is concentric, the cam is located by it in the grinding fixture as shown. After the fixture has been mounted on the faceplate of the machine, the gripping surfaces of the two jaws at the right are ground out by the internal grinding attachment, to the radius of the cydindrical dwell of the cam. The cam is clamped against the surface thus prepared, by the lever, which forces a wedge across and down upon the cam, holding it firmly into the corner in both directions.

It will be seen that this car does not employ the integral cam-shaft. By giving careful attention to the locating of the cams on the shaft and by being careful to obtain a strong drive fit between them, the difficulties of loosening and dislocation, which the integral construction is expected to cure, have been avoided. It is thus permitted to cut the cams in a way which gives the best chance for producing accurate shapes and smooth finish. The obvious scheme shown in the sketch, Fig. 19, is followed, the operation being performed on a profiling machine. The connection between the forming cam and the work is so close that the difficulties of springing and chattering, met with in the construction of the more elaborate machines required for integral cam-shafts, are avoided.

Another faceplate fixture for internal grinding is shown in Fig. 20, where it is employed for grinding the hole in the hardened nickel steel sockets used for the universal joints (see Fig. 7, Chapter I). The socket is held in the same way as when in use, by a nut screwed onto its threaded shank. It is also located in the same way, a pin in the fixture engaging a slot in the flange as shown. A limit of 0.0005

Fig. 20. Grinding the Holes in the Universal Joint Pivots

inch only is permitted in this operation, and an allowance of about 0.003 inch for the depth of the hole is the maximum, just enough being permitted for proper lubrication by the grease supply provided. This fixture is kept in place on the machine practically throughout the season. If at any time it is necessary to remove it, however, it can again be trued up by clamping a model socket in place, inserting a plug in the ground hole, and truing up the plug. These studs are held in the same way in the screw machine for roughing out the hole preparatory to grinding. The form of internal grinding spindle used should be noted. One of them is shown detached in Fig. 18, lying on the table of the machine. These spindles and their bearings are self-contained, interchangeable and adapted to work in holes of various sizes. The clutch drive provided rotates the spindle without side pressure on the bearings.

Machining the Members of the Squared Drive

As previously mentioned, the use of keys is eliminated in the drive of the Stevens-Duryea machine, their place being taken by square sockets throughout. A tapered square drive is used to connect the crank-shaft with the driving member of the clutch. The method of machining this is shown in Fig. 21. It has been found advisable to keep the milling machine set up for this work, continuously, owing to the difficulty of making a good taper square fit. When the machine has once been set, it is kept so throughout the season. An ordinary dividing head is used, as shown, tipped up to the angle of the taper. To the faceplate of this dividing head is clamped the fly-wheel flange of the crank-shaft. The outer end of the crank-shaft is supported in a suitable steady-rest as shown. For shorter lengths of crank, filling pieces are employed, having flanges bolted to the faceplate at one end, and to the work at the other. The use of filling pieces permits machining of the full line of crank-shafts without disturbing the adjustments.

Fig. 21. A Vertical Milling Machine set up for Milling the Tapered Square Drive on the Crank-shaft

The automatic cross-feed is employed in feeding the work past the end mill in the vertical milling attachment. The table has to be so far overhung that an out-board support is provided as shown, which permits this cross-feed. This consists of a sliding guide, supported by two standards, reaching to the floor and provided with jack screw adjustments for careful leveling.

The squared holes of the drive are finished on a La Pointe broaching machine in the usual manner. The further machine shown in Fig. 23 is engaged in finishing taper square holes in the clutch driving flange, this being the member into which the taper squared end of the crank-shaft shown in Fig. 21 fits. The hole is first reamed out to a taper a little larger than the distance across the flat of the finished hole. The work is then mounted on a broaching machine on the fixture shown in place. As may be seen, the broach cuts one corner of the square hole, and one-half way up each of the two adjacent sides, into the relief formed by the taper hole. A dog is fastened to the hub of the work, and the latter is mounted on a taper plug fitting the hole, with the tail of the dog located by a pin in the faceplate of the fixture, the latter being mounted on the faceplate of the machine at an angle as shown, to agree with the angle of the corner of the tapered sides.

One pass of the broach finishes one corner of the tapered hole. The broach is then returned to the starting position, the work is drawn off the taper plug, the dog indexed to the second pin on the faceplate, the work is put in position and the second corner broached. This operation is repeated until the four corners have been machined, and the square hole finished, the work being centered on the taper plug of the fixture throughout the whole operation. A taper square gage is shown lying on top of the broach in the engraving. This is used for testing the fit of the holes and the accuracy of the work, and a most accurate fit is made on this by no means easy operation. In the machine in the foreground, another operation is being done—that of

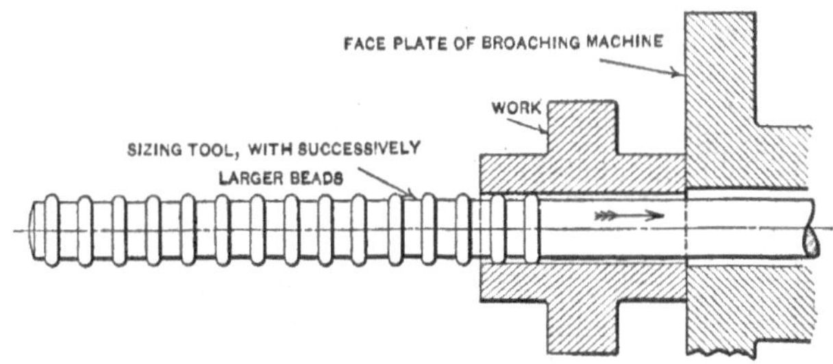

Fig. 22. Method of Sizing Phosphor-bronze in the Broaching Machine by Compression

broaching the driving slots in the driving clutch members for the multiple disks.

Sizing Round Holes in the Broaching Machine

Another unusual operation for which the broaching machine is here used, is that of sizing holes in hard phosphor-bronze bushings. This material, as any mechanic who has had any experience with it knows, is as hard on a finishing reamer as anything well can be. It is tough, elastic and slippery, and the less there is to ream the more difficult becomes the operation. Instead of reaming such holes, the tools shown in Fig. 22 are used in this shop. It will at once be seen that the operation is that of compressing the metal in the sides of the hole, until it has been enlarged to the finished size. The tool is drawn through the work. Each of the rounded rings or beads is a little larger than its predecessor, thus gradually compressing the metal the desired amount. The finished hole springs to a size smaller by some few thousandths than the diameter of the largest ring on the tool, so that the size of the latter has to be determined by experiment. This allowance varies slightly also, as may be imagined, with the thickness of the wall of metal being pressed. In such a part as that shown in Fig. 22, for instance, after drawing through the sizing tool in the broaching machine, it will be found that the hole will be somewhat larger in the large diameter of the work than in the hubs. It has been found that this difference in size can be practically avoided by passing the sizing tool through the work three or four times. Few pieces of this kind are found, however. The operation is a rapid one as compared with reaming.

An Adaptable Lapping Machine

The machine shown in Fig. 24 was built mainly in the factory, use being made, however, of the adjustable columns of a Taylor & Fenn

Fig. 23. A Set of Interesting Broaching Operations

Fig. 24. Machine for Circular and Square Lapping Operations

sensitive drill press. This special machine is intended for lapping out the square holes of the drive, but is provided also with a rotary movement in addition to the vertical movement thus necessary, so as to provide for cylindrical lapping as well. The driving pulley at the right gives the reciprocating motion, while the pulley at the left rotates the spindles through the medium of the regular geared speed drive. The sprocket wheels shown, driven from the right, are loose on the driving shaft, and carry eccentrics whose rods are extended to form racks engaging, through a suitable clutch connection, the pinion shafts by which the spindle quills are fed up and down. It is thus possible to give a rotating and reciprocating movement to the spindles, either together or separately.

Separating Piston Rings

Another milling operation is shown in Fig. 25. It is a common practice to make piston rings on an automatic machine specially rigged up for the purpose, separating the rings from the finished casting by means of a series of parting or cutting off tools, each of

Fig. 25. Cutting out Piston Rings in the Vertical Milling Machine

which is set a little in advance of the other so that the rings will cut off in regular succession. The parting tool, however, especially when used in severing cast iron work like this, having an eccentric bore, leaves a considerable burr. In the method of severing the rings shown here, the eccentric cylinder is first finished complete on the turret machine. Then it is mounted on an internal expansion chuck on the faceplate of the cylindrical attachment of the Becker vertical milling machine, as shown. This chuck is provided with clearance grooves for the gang of saws shown in the engraving. These are sunk into the cylinder, and then the work is rapidly revolved, cutting out the eight rings at once. The saws are permanently mounted on their arbor, with separating collars ground to the proper thickness.

Examples of Fixtures Used for Drill-press Operations

The drilling department seems unusually small, when compared with the size of the whole plant, and gives the appearance of being worked at high pressure. The large output required is evidently maintained by the universal use of highly developed jigs for all manufacturing operations. Multiple spindle drill presses are used to almost the entire exclusion of the single spindle type.

Fig. 26 is interesting as showing the development of the jig for a comparatively simple operation—that of drilling the cotter pin hole in a headed cylindrical stud. In the first apparatus employed (not shown) the stud was pushed into a hole up to its head, and held there by a lever, one piece being done at a time. This rigging had two faults. One piece at a time is held, and trouble with chips and burrs was experienced, as might be imagined. An improvement on this device is shown in the two jigs at the right, where a base with a set of V's is provided in which several of the pins may be placed, their heads being pressed up against the end of the V-block by

Fig. 26. Interesting Drill Jigs for a Simple Operation

springs. The cover being clamped down on the work, the parts are thus held for the drilling operation. This, however, was not quite easy enough to clean to suit the ideas of the tool designer, so the fixture shown at the left was used for the next tool of this kind that had to be made. Here hinged sides are used instead of springs as in the previous case. These sides fold up and press the heads of the work against the edges of the V-block. When they are turned down and the cover of the V-block is raised, the top surface of the V-block is all clear, so that the presence of chips shows inexcusable carelessness on the part of the operator. When the sides are folded up against the work and the cover is brought down, the latter, by means of wedge surfaces, presses the sides in, holding the heads of the work firmly in place and clamping them down on the V-block at the same time.

The jig shown at work in Fig. 27 is used for drilling and reaming the connecting-rod holes. It is of the "four-legged table" variety, with suitable clamps and hook bolts for taking the strain of the cut without permitting noticeable deflection and consequent inaccuracy in the work. A feature of the construction which is, perhaps, old enough, but probably new to many, is the provision made for both drilling and reaming with a fixed bushing, thus avoiding the use of slip bushings of different diameters. For drilling, the jig is used as shown in the engraving, with the work clamped beneath the plate and the jig bush-

Fig. 27. Gang Drill used in Drilling and Reaming Connecting-rod Ends

Fig. 28. An Unusual Array of Automatic Chucking Machines; Thirty-one are used in this Department

ings above, guiding the drills. For reaming, the jig is reversed and a reamer is used having a pilot, which passes through the work into the jig bushing (now on the under side of the plate) by which it is guided.

Fig. 28 shows what is by long odds the largest aggregation of automatic chucking machines the writer has ever seen. There are thirty-one of the Potter & Johnston type. Practically every turned part not made in the screw machine from the bar is produced on these machines. That old standby, the engine lathe, appears to be about the rarest machine tool in the shop.

Fig. 29 shows a section of the engine assembling room. It will be noted that machine tools are few and far between, the only ones in

Fig. 29. The Engine Assembling Department

sight being a drill press, speed lathe, and two or three grinding stands for sharpening tools. This shows that the manufacturing operations have been performed with great exactness. The question of assembly is simply one of bolting and screwing the separate parts together. The engines here shown are of the four- and six-cylinder type. The overhead trolley lines should be noted.

CHAPTER III

MANUFACTURING AUTOMOBILE EQUALIZING GEARS*

The present chapter deals with operations which do not present any especially unusual or spectacular features, yet they have a value derived from the fact that they are closely related to the operations which produce the bulk of the product of the machine shops of the country; for that reason they should attract the attention of mechanics interested in accurate and economical work. The operations for making a complete, compact machine unit—a differential or equalizing gear for automobile use, is described from beginning to end. The completeness of the job gives it a suggestive value that would not be

Fig. 30. The Equalizing Gear Complete, with Bevel Gear and Pinion

offered by a series of miscellaneous operations, however interesting. The value of this description, however, does not depend on its completeness alone, as many of the specific shop operations give evidence of a high degree of manufacturing ability.

Description of the Equalizing Gear

Figs. 30, 31 and 32 show assembled, dismantled and detail views, respectively, of an equalizing or differential gear, designed by Mr. A. A. Fuller, of the Providence Engineering Works, Providence, R. I. The determining feature of this design is the necessity for getting a maximum of strength and effectiveness in a minimum of space—coupled,

* MACHINERY, December, 1909.

of course, with reasonable cost of manufacture. This problem was attacked by scientific analysis. It was possible, without great difficulty, to obtain reasonable strength in the casing which contains the equalizing gearing. The crucial point was in the design of the equalizing gears themselves. In determining the proportions of the gears, curves were drawn showing the strength of the teeth for lay-outs of varying pitch and number of teeth, arranged to be contained within a casing of a given diameter. The strength and bearing area of the pivots, and the strength of the pinions as limited by the thickness of the shell between the bottom of the tooth and the bore, had also to be reckoned with. The tooth shapes were not confined to standard forms, but various pressure angles and heights of addendum were

Fig. 31. A Small Size of Equalizing Gear Dismantled to show Construction

investigated. By comparing the curves for various possible designs, a certain pitch, number of teeth and shape of tooth for the various gears were found for each diameter of casing, so proportioned that if any of the dimensions were changed, the mechanism became weaker instead of stronger. These proportions, worked into a design satisfactory in other particulars, have been adopted as standard, and the makers feel confident that it is impossible to enclose in the same space gears of greater strength than they are offering in the design illustrated herewith. As this confidence is based on mathematical calculations and has been further tested by many months of experience, it seems reasonable that they should hold to it.

Referring particularly to Fig. 32, the mechanism is contained within case *B* and covers *A* and *A'*. It revolves in the rear axle gear casing on ball bearings, mounted at the ends of casings *A* and *A'*, and the driving bevel gear is carried on the periphery of case *B*, to which it is clamped by hexagon-head screws *H*. The pivots *E* are riveted into the flanges of covers *A* and *A′*, three in one side and three in the other. These pivots carry pinions *F* and *F′* meshing with gears *C* and *C′*; the latter run in bronze bushings *D* and *D′* forced into the two covers, and are provided with broached square holes by which the floating wheel shafts are driven. As will be seen in Fig. 31 in connection with Fig. 32, gear *C* meshes with pinion *F′*, which also meshes with pinion *F*, the latter in turn engaging gear *C′*. Thus, when gear *C* is turned, gear *C′* is revolved in the opposite direction, and *vice versa*, thus forming a spur gear differential mechanism.

Attention may be called to some of the features which make for strength in this design. It will be seen, for instance, that the gears have teeth of special shape and of very coarse pitch and few numbers of teeth. The pinions have eight teeth and the gears sixteen each. In

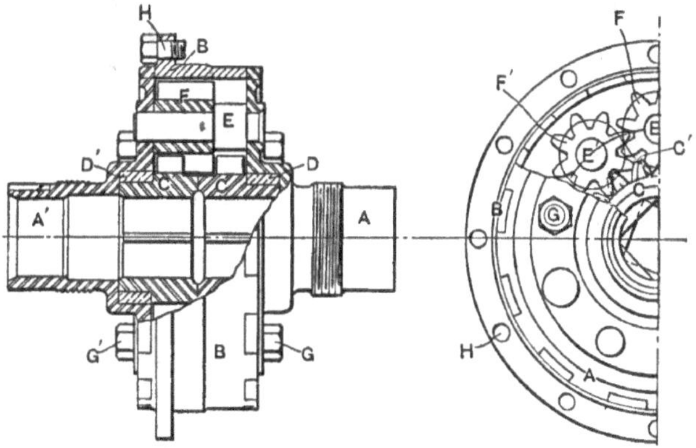

Fig. 32. Details of Construction of the 7-inch Equalizing Gear

designing the mechanism by analysis, as described, it was found that this construction was necessary for strength. Older designs of this kind, more commonly met with, in which the pinions are smaller in proportion to the gears, have repeatedly proved their weakness by breakage.

Mention should also be made of the solid way in which the parts composing the casing are fastened together. The casing *B* is provided with tongues locking into the grooves cut in covers *A*, so that the strain of transmission is taken on these interlocking members and is not taken by the bolts, dowel pins or similar parts. So far as this torsional strain is concerned, the casing is as strong as if it were made of solid metal—an impossible construction, of course. Through bolts and nuts *G* and *G′* clamp the whole casing firmly together.

The proper meshing of the bevel gears can be controlled by shifting the whole casing axially in its bearings. Nuts are mounted, for this purpose, one on the threaded diameter of *A* and the other at the same point on *A′*. By loosening one and tightening the other the teeth of the gears can be brought more closely into contact, or *vice versa*.

The provisions for oiling should be noted. The casing on the rear axle is provided with a bath of oil in which the bevel gears run. Three holes cut in the exterior of B (not shown in Fig. 31, but visible in the detail views of the operations in Fig. 33, and at the right of Fig. 34, where these holes are being drilled) admit oil from this bath into the interior spur gears. Pivots E and pinions F are grooved, as are also gears C and C' permitting a flow of oil through the whole structure, kept in constant motion through the revolving of the parts.

In describing the manufacture of this device we will take up each part in turn. The manufacture of the bevel gears will not be described

Fig. 33. Milling the Drive Tongues in the Gear Case—Second Operation

in detail, as their design is determined by the maker of the car in which the device is to be installed. The first part to be considered will be the gear case, shown at B in Fig. 32.

Operations in the Manufacture of the Gear Case

The case is made from a malleable iron casting on which the first operation, naturally, is that of snagging to remove fins, gates, etc. The second operation is performed in the Jones & Lamson flat turret lathe, of which large use is made in this shop. The casting is placed in the chuck of the machine with the flange outward. In this operation the hole is finished to size, the flange is turned, and the projecting end is faced. The regular equipment is used for this purpose, the only special tools being gages for the inside diameter of the hole and the outside diameter of the flange.

In the third operation, performed in the same machine, the part is grasped by the finished flange in special soft chuck jaws, which have been turned in place to fit the diameter they are to receive. This gives assurance that the work done in this operation will be true, within reasonable limits, with the cuts previously taken. Regular flat turret lathe equipment is used for this operation as well, suitable gages of

Fig. 34. Drilling the Three Oil Supply Holes in the Case (see Fixture at the Right), and Drilling the Bolt and Pivot Holes in the Cover

simple construction being provided. The next operation, shown at the right of Fig. 34, is drilling the three holes which admit oil to the interior of the case. This jig is of the simplest possible construction, consisting of a knee with a turned seat on which the work is placed, and an overhanging lug carrying a drill bushing. A clamp provides for holding the work, and a plug, entering a suitably located hole in the seat, provides means for indexing the second and third holes drilled, from the one previously completed. The other operation shown in this engraving will be described later on.

The tongues which interlock with the grooves in covers *A* and *A'* (see Fig. 32) have next to be milled. The fixture for doing this is shown in use in Fig. 33. It consists of a base provided with an index plate and a revolving table, by means of which the work may be indexed step by step to cut the various tongues. These are shaped by straddle mills which form the opposite sides of the tongues parallel, so that they fit into corresponding grooves milled into the covers by a straight-sided cutter. In the operation illustrated, tongues have been cut on one side of the casing, which is located in its seat in the fixture by the interlocking of these tongues with grooves provided to receive them as shown. This assures alignment of the cuts on each edge of the case. In the first operation the uncut edge of the work is simply set down onto

Fig. 35. The First Turret Lathe Operation in Finishing the Gear Case Covers

Fig. 36. Second Operation on the Flat Turret Lathe using Special Jaws

this seat. It is held down by three clamps, provided with noses which enter the three holes drilled to admit oil to the interior of the mechanism.

It is interesting to see the expertness with which the operator cuts cut these tongues. The automatic feed is set at the highest point practicable when cutting the full depth. As this would be less than the maximum possible when the cutter is entering the work, he begins with a hand-feed at a considerably higher rate, throwing in the automatic feed when the cutter gets down to work. Although the machine is of modern construction, the workman feeds in all the belt can handle. The gear casing is now complete except for certain operations performed on it in assembling, as described later.

Operations on the Gear Case Cover

The gear case covers are made from machine steel drop forgings. After the snagging, the first operation is the simple one of putting a 1½-inch hole through the center of the forgings. This is a drill press operation and is merely done to remove stock, it being, of

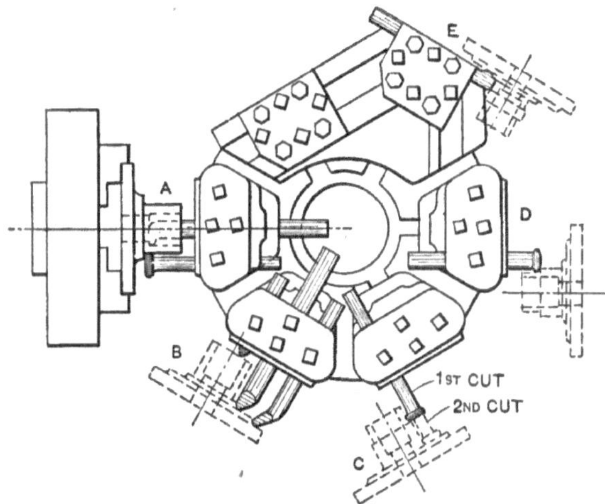

Fig. 37. Layout of Tools on the Flat Turret Lathe for the Operation shown in Fig. 35

course, impracticable to form the hole in the forging. It is next clamped by the rim with the hub projecting, in the chuck of the flat turret lathe. This first turret lathe operation is shown in Figs. 35 and 37, the latter diagram indicating the arrangement of the tools.

The first cut is shown at *A*. An outside turning and boring tool, acting in conjunction, rough turns the hub and rough bores the hole. At the next station, *B*, three tools simultaneously face the end of the hub and the two surfaces of the flange. Two cuts are taken with these, one for roughing and one for finishing. A third cut is taken with the same tools fed axially against the work to form the two grooves in the face of the flange, as most plainly shown in Fig. 32. At the third station *C*, another turning tool removes the stock on two diameters of the hub, two cuts being taken. At *D* a finishing cut is taken over the smaller diameter, while at *E* a form tool shapes that portion of the hub extending from the threaded diameter to the flange. This operation is completed in about 18 minutes.

In the second operation (see Figs. 36 and 38), the completed end of the piece is grasped in soft jaws turned to fit the surface they grasp, assuring true running of the surfaces made in the two operations. The tool at *A* bores out the large diameter of the hole, which is for clearance only. The reamer at *B* finishes the small diameter to size. The tool at *C* faces the flange, taking two cuts, one to rough out stock and the second to bring it to size. A flat-nosed tool at *D* finishes the flange. The tool at *E* roughs out the counterbore, while that at *F* finishes it. This latter tool is fed directly in, boring the diameter of the counterbore to size until the bottom is reached, when the sliding head is fed outward, so that the same tool faces the bottom of

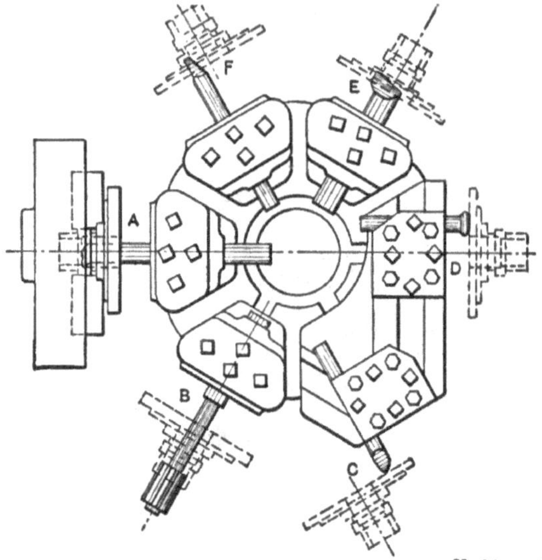

Fig. 38. Layout of Tools in the Operation shown in Fig. 36

the counterbore. The finishing is thus done by turning cuts instead of forming cuts, giving a higher degree of accuracy. Work of this kind shows the flat turret lathe to very good advantage. In the layout of tools shown in Figs. 37 and 38, there were probably no special tools of any kind required, with the exception of the form tool *E*, the rest being stock turning tools of the kind which form the regular equipment of the machine. It may have been necessary in some cases to give the tool a knock of the hammer on the blacksmith's anvil to bend it in one direction or the other, but nothing more would be needed. The cross sliding head and the multiple stops come into play in such operations as those at *B* and *C* in Fig. 37, and *F* in Fig. 38, giving each separate tool a wide range of usefulness, especially when it is so made that it can be used for both turning and facing jobs.

Of course there are all sorts of opinions about such matters, but in the question of hand *versus* automatic machines, this company believes that the conditions favor the use of the hand turret lathe in its work. The simplicity of the tooling is an important factor on contract work. The management can never be sure of the long continuance of any job, so that anything approaching costliness or elaboration is prohibited. Furthermore, it is reasonably certain that one hand machine will turn out more work than one automatic, particularly when, as in this shop, there is an inducement, such as the premium system, for the workman to get the very most out of his machine. He is constantly changing his feeds and speeds as the varying diameters, depth of the cut and condition of the tool require. He is thus able to take heavier cuts without injuring his cutting edges than would be possible without constant personal supervision.

Fig. 39. Milling the Driving Slots in a Pair of Gear Case Covers

Probably three or four changes are made in each operation to one that would be made on an automatic machine. As another advantage, this greater production of the machine means a much less capital outlay per dollar of output.

It certainly does keep the operator busy to get the most out of one of these lathes. There is no possibility of his running more than one machine, on this particular work at least. Cuts are taken very rapidly and changes of feed and speed follow each other in constant succession. There is a line of demarkation at the point where the intensity of production on the part of the hand machine and the lower capital charge on machines, buildings, stock, etc., balance the higher output per man and the consequent lessened labor cost for the automatic machines. In accordance with their judgment, some shop managers will draw the line at one point and some at another. It is fortunate for the builders of both types that all men do not come to the same conclusion when reasoning from the same premises.

In Fig. 39 the milling machine is shown rigged up to cut the driving slots in a pair of the gear case covers. The two are mounted together face to face on a special iron arbor, having a driving tail cast integrally with it in place of the usual separate dog. A formed cutter is used which shapes the bottom of the slot to the true radius of the inside diameter of the casing *B* (see Figs. 32 and 33) assuring a tight fit. This operation and that shown in Fig. 33 have to be done to close limits with good indexing plates, only 0.001 inch variation being allowed on the thickness of the slot and the tongue. This means that in order to make a good fit the dividing must be very accurate. In the cases the writer has seen assembled, these parts drove together with a very little gentle urging from a lead hammer. Not much of

Fig. 40. Jig for Drilling the Bolt and Pivot Holes in the Gear Case Covers. Another Jig for the Same Operation is shown at the Left of Fig. 34

anything else seemed to be required. In Fig. 40 is shown a jig for drilling the bolt and pivot holes in the gear covers. It is of simple construction, the cover being supported on four legs and located by a central spindle over which it is dropped and by which it is clamped, an open side collar and nut being used as shown. The bushing plate set over the work is located to bring the holes in right relation with the slots, by a tongue entering the latter. In the next operation the covers are mounted on a special faceplate, as shown in Fig. 41. This faceplate is surfaced true in place and is provided with an expansion mandrel centered integrally with it. The gear case is slipped on over this mandrel and tightened in place by turning on a wedge screw. While thus held the countersink in the outer end of the hub, the seat for the ball bearing, and the threaded diameter are turned. The thread is also cut. This is done by the Rivett-Dock threading tool, shown in operation. These operations of countersinking, turning and threading, altogether, average about eight minutes time for each piece. When the turning was in progress, the writer timed the lathe and found it was making 250 revolutions per minute, which gives about 150 surface feet per minúte for the cutting speed.

A fixture and mill of obvious construction are used for cutting the keyway by which the inner race of the ball-bearing is made fast to the hub.

Equalizing Pinions, Studs and Gears

Studs *E*, Fig. 32, are made on the Gridley automatic turret lathe with the regular tools and equipment, the job being, of course, one of the everyday variety for this machine. Oil grooves are milled, and then the burrs are removed by hand. The equalizing pinions are drilled, reamed and turned on the flat turret lathe. The ends are squared accurately to length in the engine lathe.

Fig. 41. Threading the Gear Case Covers with a Rivett-Dock Threading Tool

The equalizing gears are cut off to length from the bar stock (all gears and pinions are made of chrome-nickel steel) and are bored, reamed, faced and filleted at the large end in the Jones & Lamson machine. The hole is reamed accurately to size so as to furnish a guide for the broach in forming a square hole. This is done on the La Pointe machine at a single pass of the broach, which is a long one, having some 24 inches or thereabouts of cutting length. The outside surfaces of the gear are then rough turned on a square expansion chuck somewhat similar to that shown in Fig. 41 for the gear case cover, except, of course, that it is mounted on a square surface instead of a round one. In the next operation it is finish turned all over.

The spur gears and pinions are cut in a triple head indexing device which is one of the standard attachments on the Brown & Sharpe milling machine. Three cutters operate on three gangs of work simultaneously. By giving special shapes to the gears and by being very careful, both in centering the cutters and setting them to the proper depth, first-class results have been obtained—better than are needed in fact, since normally these gears are stationary or nearly so, being in operation only when rounding corners, in the case of a deflated tire on one side, or the slipping of a wheel in the mud. After removing the burrs by file and reamer, the gears and pinions are hardened by the regular process recommended by the makers of the steel (the Carpenter Steel Co.), with such modifications as the blacksmith of the shop has found advisable.

The equalizing gear bushings *D* and *D'*, Fig. 32, are cut from a bronze bar in the flat turret lathe, being turned and bored complete to size. A stack of them are placed on the Mitts & Merrill keyseater for cutting the internal oil grooves. The radial oil groove is cut on the

Fig. 42. A Special Fixture for Cutting Oil Grooves in the Equalizing Gear Bushing

interesting tool shown in Fig. 42. This device is a modification of the principle used in attachments for slotting screws with a saw held in the speed lathe. The knurled handle shown controls three motions. By screwing it in or out the bushing is tightened or released in the jaws by which it is held. Tripping it up or down drops the bushing away from or brings it up toward the revolving cutter, while springing it to one side brings the bushing out from under the cutter where it can be removed without interference. A wire finger locates the work with relation to the internal groove previously cut.

Assembling

The operation of assembling the parts to make the complete mechanism includes some operations worthy of notice. In Fig. 43 is a case assembled with its two covers, and dropped into a cast-iron reaming stand, where it is held from revolving by the projecting pin shown, which enters one of the three holes in its periphery. A line reamer is used, giving assurance that the two bearings in each cover will be true with each other. After this line reaming the covers are marked, numbered and burred so that the same parts will be reassembled together.

Studs *E* are next riveted to the covers, three on one side and three on the other, a hand hammer being used for this purpose. The ends of the rivets are cupped to facilitate this operation. The pinions are assembled on the studs, three on each side. The bushings are pressed

Fig. 43. Line-reaming the Pivot Holes in the Assembled Gear Cases and Covers

into the covers under the arbor press, and burred. The equalizing gears *C* and *C'* are dropped into place and the whole structure is then assembled. A square wrench inserted through the bore into the squared hole in *C*, permits the gears to turn until they are all engaged. Three bolts and nuts *G* and *G'* are now passed through, binding the whole solidly together.

It is of extreme importance in the quiet running of an automobile that the bevel gears run true. For this purpose the bevel gear seat on the outside diameter of the casing is not finish turned until it has been assembled as described. To do this, the mechanism is mounted on the lathe on large centers, bearing on the countersinks in *A* and *A'*. These countersinks, being formed in the same operation with the ball bearing seats and the threads, are true with them. After this turning and facing, a jig fitting on this accurate seat is used for drilling the flange holes through which screws *H* pass to fasten the bevel gear to the casing.

The gear is pressed into place in its seat by a simple contrivance which illustrates the demand for conveniences created by the prem-

Fig. 44. A Convenient Fixture for Assembling the Gears on the Gear Case

ium system. On the bench in front of the workman is a cast-iron seat (Fig. 44) in which the bevel gear is placed face downward. The complete differential mechanism is then placed over the gear in a position to be forced down into it. The workman now reaches up above his head and brings down the hand-wheel, clamping screw and clamp shown, which is suspended by a counterweight so as to move freely up and down and remain stationary in any position. Entering the screw in the nut in the base of the device and turning the hand-wheel, forces the casing down into the gear and thus completes the assembling. The tap bolts are now put in and are wired through holes drilled through their heads, to prevent them from turning. This completes the making of the equalizing gear.

A Good Tapping Record

While the making of the bevel gear has not been described, it will not do to pass over one of the operations met with. This is the operation of tapping the holes by which the gear is held to the flange. These holes are 5-16 inch in diameter and 13-16 inch deep and are

Fig. 45. A Tapping Operation and Operator with a Remarkable Record—75,000 Blind 5-16-inch Holes in Chrome-nickel Steel without breaking a Tap

blind, being tapped to a bottom and not through. The tapping is done in a Cincinnati drill press (Fig. 45), using an Errington friction chuck. Tapping in chrome-nickel steel by power is, it will be agreed, no "fool of a job." One of the difficulties met with is the tendency of the metal to seize the tap and break it when backing out.

The operator shown broke many taps in becoming familiar with his job, but since he has gotten into the swing of it, he has tapped 75,000 of these blind holes in chrome-nickel steel without breaking a tap.

The credit of this record must be divided between the man, the machine, the chuck and the tap, but there is enough to make a respectable showing for all four. The operator's increase of efficiency was obtained with practically no change in the tools or methods, being due simply to the training of his judgment in the feeling of the tap, and in the use of excellent tools. It might be said that a firm of the highest

Fig. 46. A Completed Equalizing Gear Set up for Testing to Destruction

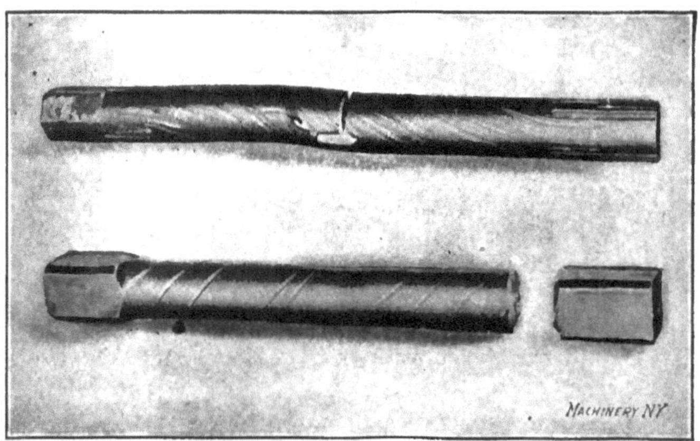

Fig. 47. Condition of Shafts Broken in Tests shown in Fig. 46; the Gears were Uninjured

reputation for accuracy and for skill in manufacturing had asked ten cents a hole for the job. This operator runs two taps in each of the twelve holes in a gear, twenty-four holes in all, in from 15 to 18 minutes.

Tests on the Finished Casings

Of course, the object that was aimed at in designing these equalizing gears for sale to manufacturers of automobiles, was to give them such strength that some other part of the machine would break first. In order to find out whether or no this result had been obtained a number of tests were made in the laboratory of the engineering school of Brown University. In Fig. 46 the casing is shown as mounted in brackets for a torsion test, the power being applied through 1-inch, 3½ per cent nickel-steel shafts, specially treated. These failed at 20,300 inch-pounds, twisting through 800 degrees before rupture. Samples of broken shafts are shown in Fig. 47, and give some idea, in combination with the figures just given, of the excellence of the material used in these shafts. No damage of any kind was found inside the gear casing, the mechanism being unbroken and running as easily and smoothly as before.